Farm Stock Library

LINCOLNSHIRE
& ITS
FARM STOCK

OTHER TITLES

The publishers have a wide range of rural titles and specialize in books covering all aspects of Poultry breeds and management. Catalogues on request (large SAE please).

LINCOLNSHIRE

&

ITS

FARM STOCK

BRIAN PADLEY

Beech Publishing House

7 Station Yard

Elsted Marsh

MIDHURST

West Sussex GU29 0JT

ISBN 978-1-85736-539-9

First published 2007

British Library Cataloguing-in-Publication Data
A catalogue record for this book is available from the British Library.

Beech Publishing House
Station Yard
Elsted Marsh
Midhurst
West Sussex GU29 OJT

CONTENTS

PREFACE & ACKNOWLEDGEMENTS

Lincolnshire has for centuries been at the leading edge of agricultural development with some of the earlier agricultural implements being produced in the county, where some of the plant breeding has taken place.

Since Roman times, thousands of acres of land has been reclaimed from the sea. It is only natural that Lincolnshire's indigenous livestock should develop along with the changing times. It has taken a lot of foresight and dedication on the part of breeders to achieve the standard which was required.

My thanks are offered to the many people and organizations who gave assistance, including the authors who are no longer with us, but have left records. Some of these are as follows:

Mr Alan Stennett who wrote the historical section.

Lincoln Longwool Sheep Breeders Association,

Lincolnshire Show Ground.

Mr Brian Sands, Developer of the Lincolnshire Buff Fowl.

The Lincolnshire Library.

Lincoln Folk Museum, Lincoln.

Church Farm Museum, Skegness.

Lincolnshire Pride Magazine, Boston, Lincs.

Lincolnshire Standard Newspapers, Skegness.

Rand Farm Park, Wragby.

References to various records and books have been made and where possible, with no copyright restrictions, have been used. I apologize if I have omitted reference to any source, which would not be intentional.

Brian Padley

FOREWORD & POEM

LINCOLNSHIRE MY COUNTY

by Gladys Waite

A "yellow belly"* I'm proud to be, to live in a county next to
the sea.
The folk who live here are hardy and bold, for the North-easters
blow so bitterly cold.

When people tell me it's dull and flat, it makes me mad it's not
like that:
For beauty all around I see, the sunset, skies, each flower and
tree.

We stood together throughout the war, each night we heard the
Lancasters roar,
When morning broke we would hear them drone, as some of
them struggled hard for home.

Together we toiled in fifty-three when the coast was invaded by
a terrible sea.
So great was many peoples' plight we never never can forget
that night.

Our county seems to be the tops at growing really first class
crops,
For quality, farmers have quite a name, but they all grumble just
the same.

In spring the fields are green and fine, everything growing in a
dead straight line,
Come May, tulips and daffs by the acre grow, the Spalding pa-
rade is a marvellous show.

Potatoes grow in that rich dark soil in great demand for chips or
to boil,
We've fat juicy sausage and tasty stuff chine, our local foods

are really sublime.

Lincolnshire bred some famous men, Smith, Banks, Flinders and John Franklin,
They risked their lives and crossed the sea, to bring back knowledge for you and me.

Tennyson's poems world wide are read, an apple fell on Newton's head.
Brave men drained the fens at tremendous cost and in that mud King John's jewels were lost.

Of historical buildings we've quite a few, Louth Church, Boston Stump, are but just two.
Lincoln Cathedral on a hill stands high, Tattershal Castle, Crowland Abbey, gaunt to the sky.

Our animals, Curly Coat pigs produced lovely sweet pork, mighty shire horses did most of the work.
Long wool sheep warm cardigans make, and Lincoln Red cattle, you can't beat for steak.

What variety our county holds, the rivers, the shore, the marsh, the wolds.
Out on a limb we seem to be, but from serious violence we are free.

We're really quite a happy lot, contented with the life we've got,
Living in peace twixt wold and sea, is there anywhere else you'd rather be.

---

* Editorial Note: Yellow belly comes from reference to the yellow waistcoat of part of the uniform of the Lincolnshire Yeomanry.

Also may be references to the frogs with yellow bellies that lived in the marshes and fens before they were drained.

1

LINCOLNSHIRE
AN
AGRICULTURAL COUNTY

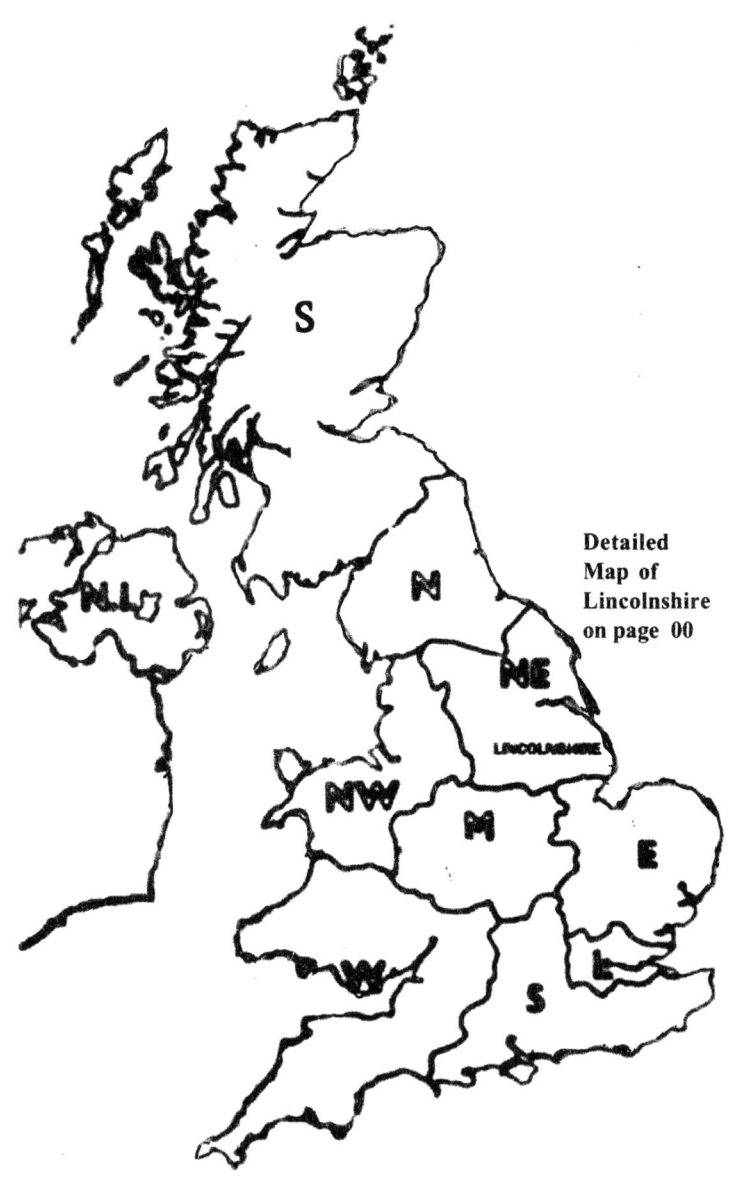

Detailed
Map of
Lincolnshire
on page 00

Areas of United Kingdom

LINCOLNSHIRE AND ITS FARMING

The county of Lincolnshire can claim, with some justification, to be one of England's premier agricultural areas. In fact, those of us who know it well would describe it as *the* premier agricultural county.

Location

Set on the East Coast of England, between the two great water barriers of the Humber and the Wash, and bounded on the west by the Trent, Lincolnshire has sometimes seemed to be somewhat cut off geographically, but it has always been in the forefront of farming in the UK.

A Farming County

Agricultural activity can be traced here back into the Iron age, and possibly earlier, with much farming taking place during the Roman period, when it is believed that grain grown on the lowland areas was shipped up the Witham and along the Foss Dyke Canal, one of the earliest man-made waterways in Britain, to the cities and troops in the north.

In Norman times, the Domesday Book shows that Lincolnshire had the highest proportion of what in later years became the yeoman farmers of England — men farming their

---

* This historical section provided by Mr Alan Stennett, Woodhall Junction, Kirkstead Bridge.

own plots of land, and a tendency to be independent though an action which has been both a blessing and a curse to local farming over the years.

During the Agricultural Revolution, Lincolnshire was an enthusiastic follower of trends, but it didn't really start to set them until the great agricultural machinery firms of Ruston, Robey, Marshalls, William Foster, Clayton and Shuttleworth and many others helped the process of mechanisation that has turned Lincolnshire into the great arable county that we see today.

The preponderance of arable farming today should not, however, obscure the fact that, as Brian has been finding, Lincolnshire has a great livestock tradition although, sadly, most of the breeds with which it has been most closely associated are now classified as 'Rare' since they no longer match the needs of the modern farmer or consumer.

Farm Stock

The Lincolnshire Longwool sheep was once the mainstay of the principal export trade of the county, and much of the country, producing, as it did, large amounts of that most valuable commodity before the advent of synthetic fibres -- wool.

Huge number of longwool sheep were kept on the Wolds, fens and marshes, and Lincoln and Boston both held the title of 'Staple' towns - able to export wool to the weavers and traders overseas

Lincoln Red cattle are still with us, although in much reduced numbers, but their good mothering ability and kind temperament still make them a popular suckler cow.

Sadly, we have lost our native pig breed, the Curly-coat, and despite rumours that there might still be small pockets holding out in places like Hungary, none has ever been found.

There has never been a native Lincolnshire horse breed, although the Great Black Horse of the fens, a huge animal bred to lug loads through soft ground, is generally credited as being one of the main forbears of the Shire, developed just west of the county in the Leicester area.

The story of farming in Lincolnshire, and especially the livestock side is a long and complicated one, and I commend Brian for his efforts to draw together elements of this fascinating tale.

POSSIBLE REASONS FOR DEVELOPMENTS

As noted, Lincolnshire is primarily an agricultural county, where the population have to live from the land. Much is flat, marshy and cold, with the wind blowing from the sea.

The farmers have to suffer great hardship and have had to adapt the livestock to suit the environment. Pigs and sheep were developed with long woolly coats so they could bear the hardship. The cattle have gone through many changes to meet the needs of the market. On the poultry side, the Lincolnshire Buff Fowl has been revived and is doing nicely, and it is a mystery why it was allowed to become extinct.

There has been *adaptation* at many stages, and this recognizes the very nature of the animals being kept. Charles Darwin recognized the inbred tendency for some breeds to be more suited to specific conditions. He found, in fact, that animals would differentiate and accept the conditions which suited them best.

Thus it seems has been the case in Lincolnshire. He stated:

> Different races of sheep, like cattle, present constitutional differences. Thus the improved breeds arrive at maturity at an early age, as has been well shown by Mr. Simonds through their early average period of dentition. The several races have become adapted to different kinds of

* *The Variations of Animals & Plants Under Domestication*, **Charles Darwin, London, 1883**

pasture and climate: for instance, no one can rear Leicester sheep on mountainous regions, where Cheviots flourish.

As Youatt has remarked, " In all the different districts of Great Britain we find various breeds of sheep beautifully adapted to the locality which they occupy. No one knows their origin; they are indigenous to the soil, climate, pasturage, and the locality on which they graze; they seem to have been formed for it and by it."

Marshall relates that a flock of heavy Lincolnshire and light Norfolk sheep which had been bred together in a large sheep-walk, part of which was low, rich, and moist, and another part high and dry, with benty grass, when turned out, regularly separated from each other; the heavy sheep drawing off to the rich soil, and the lighter sheep to their own soil; so that " whilst there was plenty of grass the two breeds kept themselves as distinct as rooks and pigeons. " (Autor's bold italics).

Numerous sheep from various parts of the world have been brought during a long course of years to tho Zoological Gardens of London; but as Youatt, who attended the animals as a veterinary surgeon, remarks, " few or none die of the rot, but they are phthisical (consumptive); not one of them from a torrid climate lasts out the second year, and when they die their lungs are tuberculated."

There is very good evidence that English breeds of sheep will not succeed in France. Even in certain parts of England it has been found impossible to keep certain breeds of sheep.

Champion *McTurk Trojan* with Graham Lowry
Royal Highland Show

Champion *Brandon Nero* led by Mrs J Burtt
Lincolnshire Show

LINCOLN RED CHAMPIONS

COLONY CARR HAZEL Winner; owned by H M Prison

Line-up of Lincolnshire Sheep at a Show

LINCOLN RED BULL Painted by Albert Clark, c. 1910
Museum of Lincolnshire Life .

CHAMPION: CONEY'S LONGWOOL LINCOLN YEARLING EWE

PLATE 36 PUNTING FOR WATERFOWL

PLATE 37   SLEDGING FOR WILDFOWL

Old Pastimes: From ***Old Sports of Great Britain,*** Hy Alken, BPH

PLATE 43 ANGLERS

PLATE 32 WILDFOWL SHOOTING

Old Pastimes: From *Old Sports of Great Britain,* Hy Alken, BPH

Linconshire Buff Pair (Brian Sands)

Linconshire Buff Pullet (Lucy Hampstead)

H M PRISONS COLONY CARR HERD
Courtesy Mr Alan Stennett of Woodhall

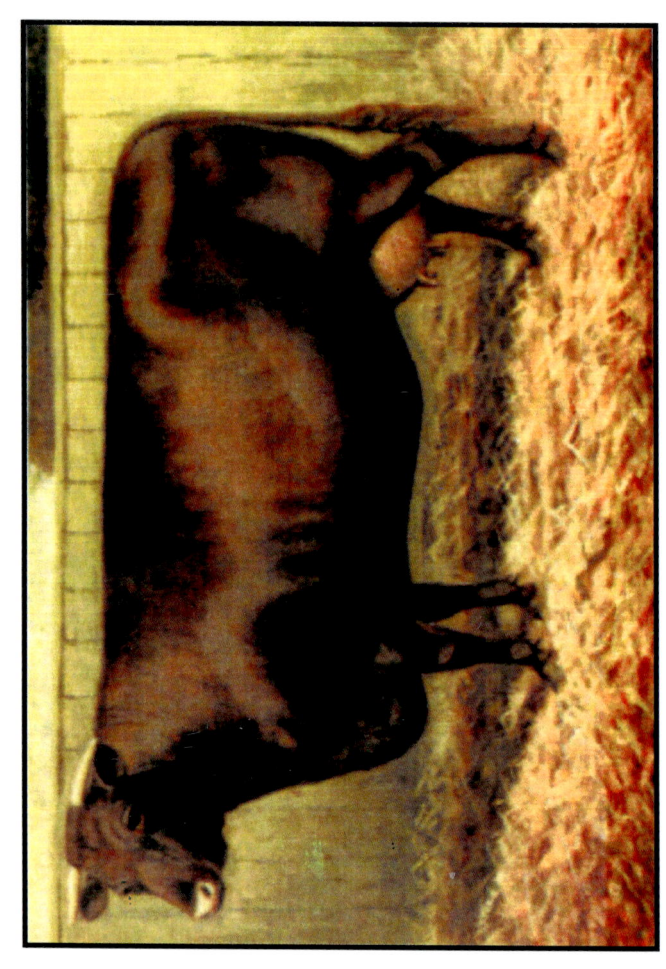

LINCOLN RED COW Painted by Albert Clark, c. 1910
Museum of Lincolnshire Life

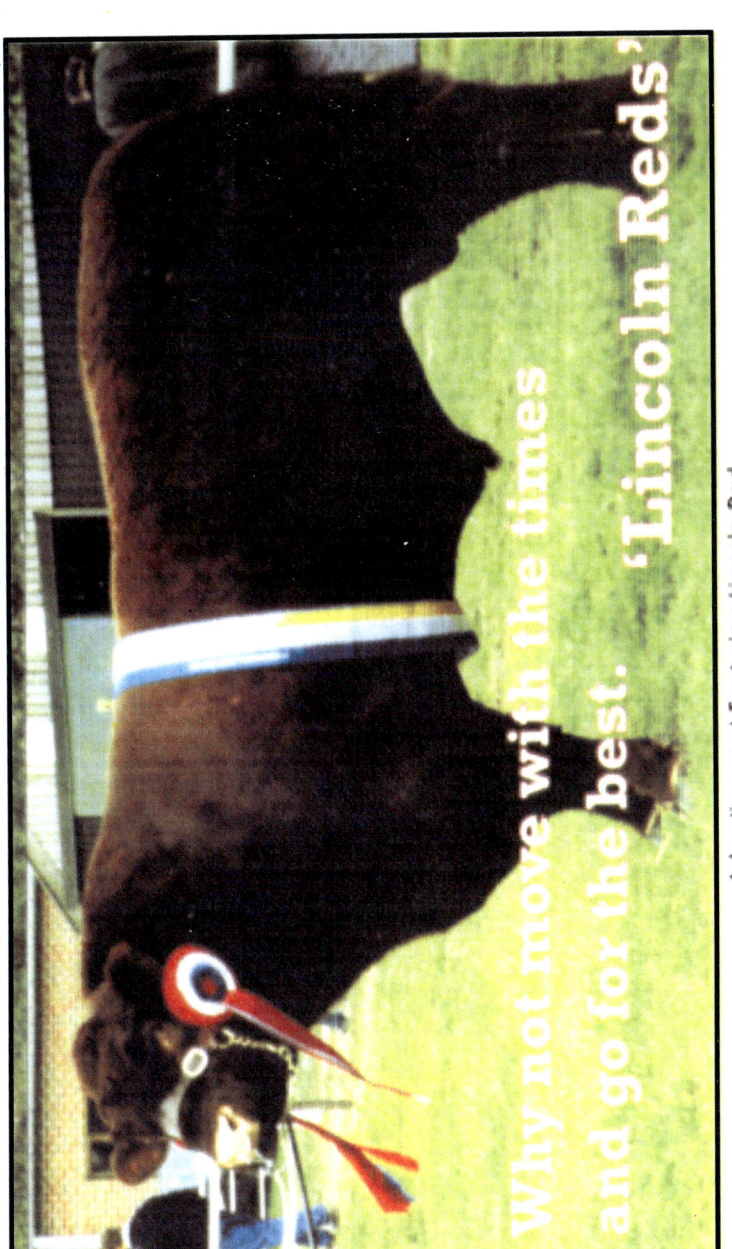

Why not move with the times and go for the best. 'Lincoln Reds'

Advertisement Featuring Lincoln Red

'PATRIOT' THE VALUED SHORT HORN BULL, 1810
Based on a painting by William Ward

2

LINCOLNSHIRE
AS A
THRIVING
COUNTY

ARMS OF LINCOLN

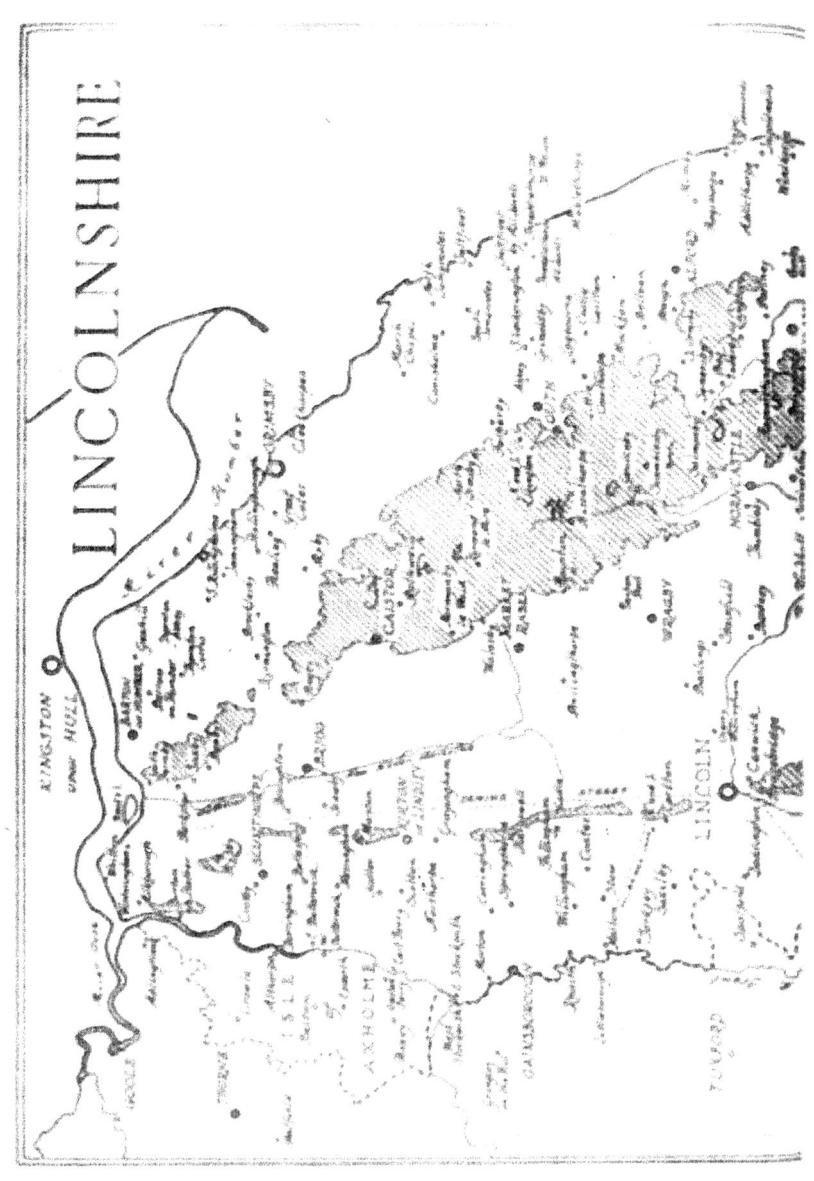

Map Showing Principal Places

Map (2) Showing Principal Places

THE LOCATION

The county measures around 75 miles long and 45 miles deep and is situated on the East Coast between the Humber and the Wash. One side is fringed by the North Sea with its bitterly cold winds, making winters very harsh and not a place for weaklings.

The people therefore are strong and what is regarded as 'hard headed' willing to work hard, but also with the idea of profit at the end. One writer has described them as 'uncompromising, dour in parts, ugly in patches, downright, and of thorough-going English independence'. *

It is a county of tradition and ancient history, which included the Saxons that were prepared to defend England from the many invaders.

A DIVERSE COUNTY

Because of the wide diversity there can be no generalizations. It is a large area made up of many different segments which comprises:

1. The ancient city of Lincoln with its cathedral which dates back to 1078. It is situated on a high plateau with the cathedral at the peak.

There are also old churches and other buildings of architectural

Cont. p23

J Wentworth Day,in *Lincolnshire,* **from** *The English Counties* (*ibid*)

HIGH BRIDGE LINCOLN

There are many ancient buildings in the city, this being one.

LINCOLN FROM WITHAM BANK

The site of Lincoln is not readily surpassed on the Eastern flamk of England. The river Witham in the lapse of ages has carved in a limestone-capped plateau a valley which at this point is rapidly broadening out towards the great tract of fen which intervenes between the wolds and the sea.

Here at the edge of the plateau, the Romans built the vallum of a military station, which they called *Lindum Colonia*, and thus laid the foundation of the city of Lincoln.

From: *Our Country,* Cassell & Co, London, nd

interest.

Moreover, this is a thriving bustling city where there is much commerce and the shopping areas to meet the demands from outside.

2. Many towns and villages, again, with a long and distinguished history.

There is Grimsby noted for its fishing port and related industry. Scunthorpe at one time famous for its steel industry, although now in decline.

Other towns include Louth, Grantham, Gainsborough, Stamford (famous for the old sport of Bull Running), Boston, Spalding (more details later) and seaside resorts which include Cleethorpes and Skegness, popular for those who wish to holiday on the east coast.

3. The farming country which is composed of the marsh and fens worked upon and rescued from the bogged land and sea floods to become very prolific growing areas, and the wolds which rise high above the marshes and contain houses, farms and villages.

The Fens

The bogs of the fens are now rich farm lands, which command very high prices. It was the home of the duck decoys, some 38 were in existence, where thosands of wild ducks were captured for the market. They were in the districts around Sleaford and Wainfleet, but have now succunbed to the march of progress with the agricultural revolution which has taken place.

Duck Decoys in Action

LINCOLN CATHEDRAL FROM THE SOUTH WEST

LINCOLN & ITS CATHEDRAL

Lincoln and its cathedral are steeped in history and have been associated with many great men. Some idea can be obtaned from the following*:

> Lincoln has, then, an especial interest as being; "one of the first important buildings erected wholly in the Pointed

From: *Our Country*, Cassell & Co, London, nd

style. Though less complete in some respects than Salisbury-(begun in 1220 and finished in 1258), it was commenced full thirty years earlier.

Some critics have asserted that the architecture of Lincoln exhibits signs of a French influence; but the late M. Viollet-le-Due, whose authority on this point scarcely admits of dispute has declared that, after the most- careful examination, he could not find in any part of the cathedral of Lincoln, neither in the general design, not in any part of the system of architecture adopted, nor in the details of ornament, any trace of the French school of the twelfth century, so plainly a characteristic of the cathedrals of Paris, Noyon, Seulis, Chartres, Sens, and even Rouen.

This fact, which greatly increases the probability that the architect, Geoffry de Noiers was an Englishman, gives us good reason to claim for St. Hugh the distinction of having been 'the first effectual promoter, if not the actual inventor, of our national and most excellent Early English style of architecture,' and in point of interest renders it difficult for any other church to exceed Lincoln Cathedral.

In size and importance it may be regarded-as the third great church of the Early English period in England, the whole of the interior; except the presbytery, being of this age; and this part follows so immediately after the rest as not to produce any want of harmony, but merely a degree of enrichment suitable to the-increased sanctity of the altar and the localities surrounding it.

Lincoln cathedral is regarded as one of the most beautiful of British cathedrals.

THE CLOISTERS

LIVE STOCK DEVELOPMENT

Lincolnshire is renowned for its live stock much of which was developed to stand up to the harsh climate. Those breeds which are unique to the county are covered in detail in other parts of the book.

There was no special breed of horse developed which is surprising, but no doubt the great Shire horses were able to cope with any of the work. A breed of horses known as *Wildmore Tits* were kept before the fens were drained and these were allied to the Arabian breed. Thousands used to be bred and some offspring of these, crossed with other breeds, may still be seen on some farms.

Bantams were always of interest and a little known fact is that the first Black and White bantams, now known as Rosecombs were in existence prior to 1483, being kept in Lincolnshire. This was at *The Angel Inn*, Grantham, the inn-keeper being a John Buckton who left them to his son. Richard III stayed at the inn and is reputed to have admired them*.

These were probably the first bantams to be in exist-ence in Britain. Today they are highly prized as top class show birds. However, it is not a breed for the novice because to birds are difficult to breed.

They cocks have large tails with broad feathers and the head of the male is quite extraordinary. The Rose comb has a

* *The Rosecomb Bantams*, Joseph Batty, Elsted, 1998.

White Rosecomb Bantams

Fine
Workings
on Comb

Large
White
Earlobes

Head of Rosecomb Male

Very difficult to achieve the head points and large tail.

long leader (spike) and the main part is covered with fine work-ings which must be without blemish; in addition, the earlobes must be large, round and velvety in a white colour, without any marks or discolouring.

FARM ANIMALS & OTHER ACTIVITIES

The people of Lincolnshire are practical in their farming and related activities. The animals developed within the county are explained in later chapters.

There has had to be a forward looking approach and changes made when necessary. Before the fens were drained there was great activity on the waterways. Catching wildfowl was an industry in itself, occupying many people. Hides were built and special boats were used for shooting the wild ducks.

Artists like Henry Alken captured these historic scenes in paintings, examples being given in the Colour Section.

3

THE
FARMING
COMMUNITY

WOOLSTHORPE ON THE LINCS-LEICESTER BORDER
BELVOIR CASTLE IN THE BACKGROUND

A Broad Canvas

From *The English Counties (ibid)*

THE FARMERS

Recognition of the skills required and, when necessary, to change and adapt, are essential for survival. This has been present in Lincolnshire to a high degree. The county facilities, including education, have been present to assist and develop the potential.

Although skills are passed down these are generally old skills so the young coming into agriculture must adapt and become conversant with modern methods. This has been propounded by a number of educationalists, and the following is typical:*

. Of the necessity of an agricultural education, independent of his other scholastic acquirements, for the young or rising farmer at the present day, no one of the slightest intelligence has the smallest doubt.

The customs of our grandfathers are already old and effete; whilst those of our fathers are fast wearing out. In a lecture delivered by Mr. Morton, at the Royal Agricultural College, has laid down these three points as essential to the young farmer:—

1st. That he should have practical skill.
2nd. That he should have business tact.

---

* *Rural Life*, John Sherer, London, nd.

3rd. That he should have a liberal and scientific education.

If a man possessed merely practical skill, and nothing more, he was little better than a labourer; that if, added to his practical knowledge, he had business tact, he might not only be a labourer, but become the manager of a farm; but that, without having, in addition to capital, a liberal and scientific education, he was not fitted to hold a large farm in the present day.

" The more possession of capita " does not qualify a man for being a farmer. ....

An Old Scythe

An old tool which has been largely superseded by modern technology.

AGRICULTURAL SHOWS

Agricultural shows provide contact within the county, and help to maintain standards in animal breeding and management. They are also a method of marketing products and,at the same time, to let the general public see what the farmers are doing.

Within Lincolnshire many shows are organized, as would be expected in an agricultural society.

Shows & Exhibitions In The County

The live stock breeders and growers in the county attend many shows in other areas. The following are held within Lincolnshire:

1. Lincolnshire Show usually in June.

2. Woodhall Show. Held in the Spring and usually has classes for Lincoln Red Cattle and Lincoln Longwool Sheep.

3. East of England Show, Peterborough . Usually in July.

There are other specialized shows for various breeds of animals and poultry.

AN ACTIVE COUNTRYSIDE

As noted earlier, the marshes and fens have been drained and developed by farmers and smallholders, all of whom make a profitable living from the land.

The once wild and uncultivated land is now worked to the full. The lush pastures are ideal for cattle and the lesser areas produce fine sheep. Other crops include corn, potatoes, sugar beet, fruit and many types of vegetables. The sheep provide wool and lamb or mutton and the Lincoln Reds are now beef as well as milking cattle, thus supplying products from them.

Spalding has developed into the bulb and flower-plant supplying town of Britain. Intensive methods are employed to produce a wide range of plants. In addition, tomatoes are grown on a large scale.

Modern methods of farming have been developed so that the land could be ploughed deep, making it more amenable to drainage. This has been augmented by farm yard manure so the land has become rich and fertile.

Farm machinery is utilized to the full because the farmer believes in making the most of the land he occupies and works. There is no standing still because to make progress profits have to be high, thus allowing maximum investment in the machines required.

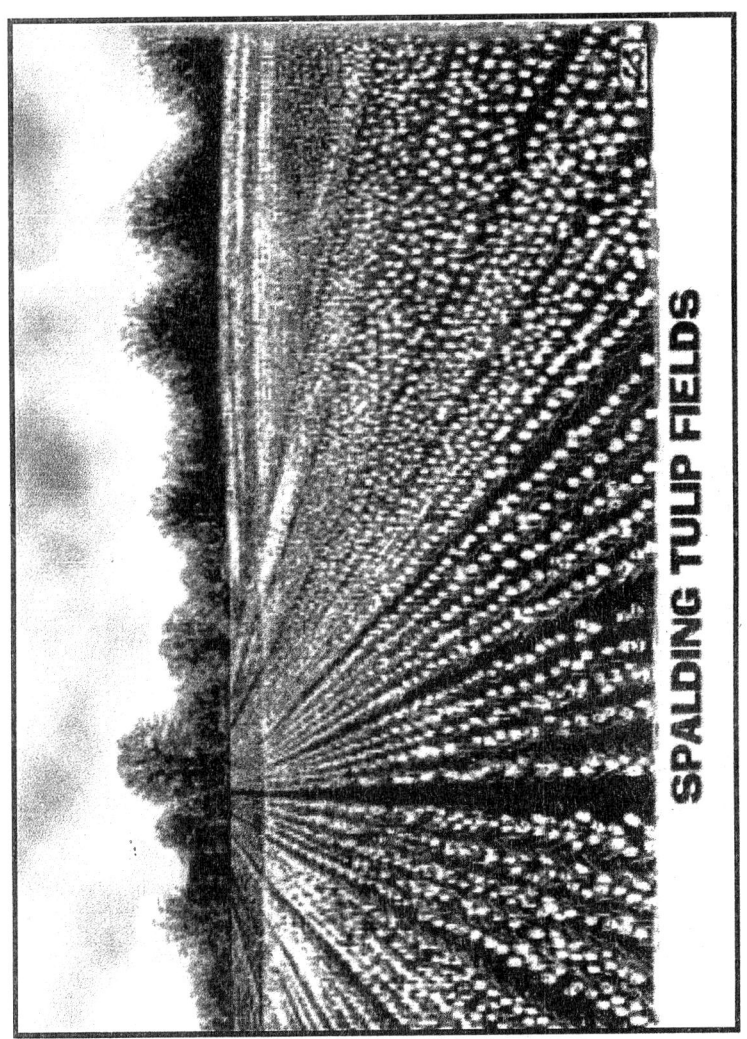

SPALDING TULIP FIELDS

PLOUGHMAN AT WORK

THE FARM STOCK

The farm stock developed in Lincolnshire are as follows:

1. Lincolnshire Curly Coat Pig

A very useful animal, but it became a colossus, specimens growing to 40 stone. Also lean bacon became the fashion. As a result the animal is no longer kept, but could be revived and modified.

2. Lincolnshire Long Wool Sheep

Very prolific wool producer and by adaptation is a very useful producer of mutton.

3. Lincoln Red Cattle

A useful, docile breed which has undergone many changes. In its original form is now rare, but the breed does still exist.

4. Lincolnshire Buff Fowl.

After being virtually extinct they have been revived and is once more being kept. Care must be taken to ensure it is not spoilt by insistence on 'fancy show points' which are likely to defeat its utility status.

LINCOLNSHIRE CURLY COAT PIG

The breed is mentioned in very early books on pig keeping and generally commended. One of these early works was ***Domestic Pigs*** by H D Richardson. published in 185 4. He states:

> The old Lincolnshire breed was light-coloured, or even white, in most specimens, a curly and woolly coat. These pigs were of medium size, were good feeders, came early to maturity, and fattened easily. This county is famed for its strain of pigs ; not only the county, but many private gentlemen having a breed to which they give their name, or the name of their properties. The improved breed , generally are white with fine skins, sparingly covered with slender bristles, ears erect and pointed, and the body long, straight, and round.

From this paragraph, written about 1850, it will be seen that the breed had been modified from its original type. Later, it also increased in size until it was no longer an economic proposition.

The decline of the breed took place over a long period. In 1947 in England and Wales there were 2,652 boars licenced for Large White Yorkshire pigs, but only 18 for Lincolnshire Curly Coats. Although not at the bottom of the rankings it was quite low down.

LINCS CURLY COATED SOW

LINCS CURLY COATED BOAR

DEVELOPMENT OF THE LINCOLN SHEEP

In a separate chapter the Lincoln Sheep are covered. They are a supreme breed which is a great accolade when it is considered that there have been around 30 separate breeds. The improved Lincoln Longwool Sheep are the native sheep of the county. As wool producers they stand pre-eminent in this country.

A concise history is in *The Book of the Farm,** a standard work on agriculture. It is acknowledged that this breed was developed in the county which bears its name and was improved by judicious crosses with the Leicester, which improved, but did not impair the wool producing characteristics. Thus:

Prior to about 1850 they were flat-sided, ungainly, slow-feeding sheep. Now, thanks to a judicious admixture of Leicester blood, and to careful breeding and management, they are vastly improved in form as well as in fattening properties; while their wool production has in no way deteriorated.

In the fairly rich pastures of their native districts no other breed can equal them as rent-paying sheep, and in recent years there has been a growing demand for Lincoln sheep, not only in other parts of England, but also in for-

---

* *Stephen's Book of the Farm*, Edinburgh & London, 1891

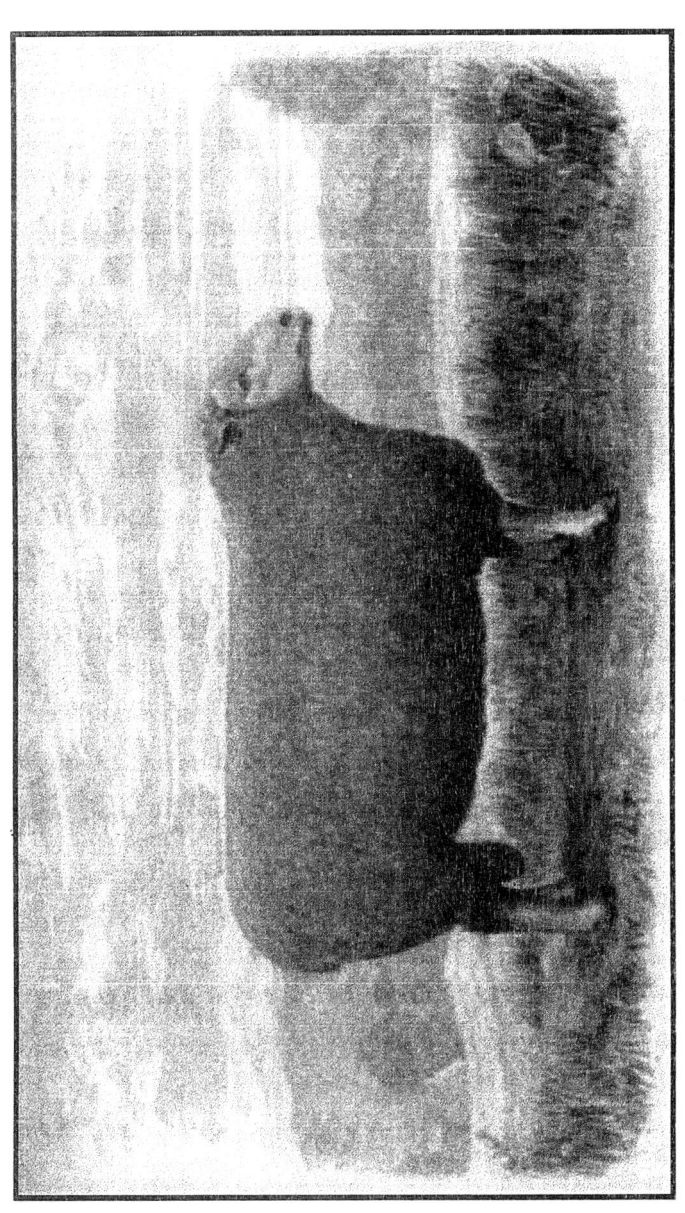

LINCOLN RAM

eign countries. They are hardy and wonderfully prolific, about 30 per cent of the ewes producing twins, triplets being frequent, and now and again four lambs are dropped at one birth.

The body of the Lincoln is smaller but more symmetrical than that of the Cotswold, the back broad and firmly flushed, the ribs well sprung, the shoulder deep and wide, neck thick, head pretty strong, with good legs of lean mutton. It matures at an early age, and the fleece, which is long and lustrous in the staple, often exceeds 20 Ib.—indeed as much as 26.50 lb. of wool has been clipped from a fourteen-month sheep.

Readers will notice that credit is given to the rich pastures of their native county. They would not necessarily yield the same results in an entirely different environment, although, they are quite versatile and flexible, as shown by their wide spread adoption overseas.

NEXT PAGE

Distribution Map of Sheep in England and Scotland. Some breeds may be found outside the areas designated.

After N L and Julia Tinley in *Good Sheep Farming* a section in *Modern Farming*, S Graham Brade Birks, London, 1950

DISTRIBUTION OF BRITISH BREEDS OF SHEEP

THE LINCOLNSHIRE RED SHORTHORN

The present-day Lincoln Red history goes back to the close of the 18th century, when Mr. Thomas Turnell, of Reasby, developed this breed, which were then known as *Turnell Reds.* So gradually a special breed of Shorthorns of distinctive character came into being, red shorthorns of a milking type.

In 1895 the Lincoln Red Shorthorn Society was formed, and the stock was registered and protected. The merits of the dairy breed are that it is a useful farmer's cow, a heavy milker and a good store producer.

Apart from dairy powers the Lincoln Reds are well known for their beef-producing qualities, and steers at less than three years old give 8 to 9 cwt. of meat.

As will be shown many changes have taken place and Polled types are now the ones preferred.

Other breeds of cattle, notably the Friesians are now kept as well.

Some of the original Red Shorthorns are kept, but often in small numbers by farmers who prefer a docile animal that can be milked by hand for local consumption.

LINCOLNSHIRE RED SHORTHORN COW

LINCOLNSHIRE RED SHORTHORN BULLOCK

LINCONSHIRE BUFF POULTRY

This breed, once extinct has been developed again and given recognition by the Poultry Club. There is also a breed club to look after its interests.

Recently a book has been published on the breed, in the belief that this fine utility breed should now be given prominence and become more popular.*

Harrison Weir** the great poultry author and artist recommended the original breed:

As it is, the Lincolnshire Buff is a good winter layer of medium sized eggs of a. variable light yellowish brown, in shape mostly of a roundish oval, though some few are elongated.

They are steady, patient incubators, and good mothers, and the chicken are easily reared and fatten readily. The flesh is juicy, delicate in flavour, and white, the skin thin, the fat white, though in some instances this is not so in the much-feathered or yellow shanked.

If well fed, fatted, and finished, they are a very saleable market fowl, and as a cross-breed can be recommended as such; but always those with clear shanks should be kept.

---

* *Lincolnshire Buff Poultry,* Joseph Batty, Elsted, 2007
** *Our Poultry and All About* Them, London, 1902/4

An Original Type of Lincolnshire Buff Male

Drawn by Harrison Weir

The Lincolnshire Fens have long been noted for good poultry, as the following, excerpt from the "Art of Longevity," by Edmund Gayton, 1659, will testify:—

> "What droves of Higlers post in from the Fens
> With Fowls most Epicene, both cocks and hens,
> Of all which company I don't enjoy,
> One Duck, and yet related to a Coy."

So even at this period it appears that the Fen fowls were both numerous and good—judging so by the second line of the verse in praise of both the cocks and hens.

4

THE
LINCOLNSHIRE
CURLY COAT
PIG

LINCOLN CURLY COATED SOW

THE LINCOLNSHIRE CURLY COAT PIG

The *recognition* of the Lincoln pig, was as late as the twentieth century is remarkable as the establishment of the large white was fifty years earlier. It's not clear whether its standard was improved by the introduction of Bloodlines of the improved swine of the neighbouring county.

The Lincolnshire Curly Coat evolved at a time when the requirement was for a large animal producing mainly bacon in great quantities, at that period many of the farms workers were boarded a the farmhouse, bacon was a major item of their diet. There were at that period much larger families. A family of 12-14 living in a cottage was not uncommon.

The Lincoln Pig became ideal as a cottager's pig with its quick growth and its capacity to convert cheaper foods such as waste potatoes and barley meal into bacon in quantity. Mature pigs could weigh up to 7cwt, its main fault being that it got too fat in relation to lean meat.

The Ideal Type was being light in head and jowl and shoulder with great length of body with moderate depth of sides.

It has been pointed out by Douglas "in the meat industry" that ten pigs competing at Smithfield in 1908 in the class for pigs between 220 and 300lb live weight averaged 215 1b at 316days old has gained 10. 9oz per day since birth.

Ten Lincoln pigs which were eligible for the same class

weighed 178 lb each at 161 days , which makes an average daily gain of 1 lb 1.68ozs.

In 1911 five pairs of Lincoln pigs between nine and twelve months old averaged 8cwt 2qr and 221 lb per pen or 553 lb per pig at the age of eleven months, or in the younger class for pigs under nine month, sold the heaviest pen of Lincoln's which scaled 7cwt 2 lb each or 393 lb each at seven months and four weeks, having established themselves as the heaviest white variety.

DESCRIPTION

The main points of the breed are as follows:

Hair white and fine not too abundant but curly or wavy, straight or coarse hair being objectionable. The skin is soft.

The head of medium length; wide between the ears , a long snout or dished face, being contrary to type. The neck is of medium length, the ears not large as compared with the size of the pig and bent over the face matching the carriage of the head, the line of the jaws being practically on a line with the belly.

The chest as in all breeds, is wide and deep, the back long and straight making a symmetrical curve. The ribs are well sprung to correspond with the size of the body, the sides are deep giving width to the flitch. The Loin is broad and strong when viewed from the side the hams are deep and thick.

Great stress is laid on the width of the shoulders and the formation of the hams The sows are prolific and have good mothering instincts, because of their size there is risk of losing some piglets by the mother overlying on them.

A summary written around 1918 stated as follows*:

LINCOLN CURLY-COATED

Until the last few years this ancient breed of large pigs was only well known in East Lincolnshire. The abundant white hair is curly, as the name indicates, and the skin is of the same colour, though some blue spots are usually present. The face is shorter than that of the Large White, and the ears (which should be of moderate length) fall over it, while the nose is straight. It is claimed to be unequalled for early maturity and development, and is undoubtedly hardy and vigorous. It crosses well with other breeds, particularly Berkshire, Large White, and Large Black.

---

* From *Elements of Agriculture*, W Fream, London, 1918.

THE WHITE PIGS

White pigs have been in existence for centuries, but there has been a merging of the breeds and selection in different areas, so they became separate 'species' and often given a name to denote the area in which they developed. The local market had a considerable influence, including the preference for the type of meat. Size was also important as was the ease of rearing and the fattening properties of the breed.

British history, so far as it concerns the pig, goes back to the time when the Book of Domesday was compiled. In it we find many a reference to pigs, particularly in regard to the acreage of woodland which in those days was the natural and proper home of the animal.

The size or value of a piece of woodland was reckoned by the number of hogs it was estimated to support, so much so that Walter of Henley, author of " Husbandry," a work that was much in vogue from the thirteenth to the sixteenth century, went so far as to suggest that the hand feeding of pigs was outside the range of practical politics:

" The swineherd ought to be on those manors where swine can be sustained and kept in the forest, or in woods or waste or in marshes, without sustenance from the grange."

A more modern chronicler.. Thorold Rogers, in his " Six Centuries of Work and Wages," has a good deal to say as to the earlier phases of pig-keeping in Great Britain . "It is to be ex-

White Leicester Pig

White Suffolk Pigs

Both breeds may have figured in the development of the Curly Coat pig.

pected," says he, " that swine were, for general use and consumption, the most important objects of all English agriculture in the thirteenth century and for many a century afterwards".

The variations based on areas are numerous: thus:

Dorset

Hampshire and American Hampshire

Essex & Neapolitan Essex

Black Suffolk

Norfolk Thin-rind

White Leicester

Lincolnshire Curly Coat

Cumberland

Wessex

Yorkshire

Chinese, which was one of the early breeds.

Undoubtedly, the Lincolnshire Curly Coat pig was developed from other white pigs, possibly the Yorkshire and White Leicester along with local stock. In some respects it was quite domesticated and therefore relatively tame because it was in great favour as a pig for the cottager.

The breed had to be hardy and that would be the reason for breeding those pigs which had thick curly coats, thus being able to withstand the harsh winters of Lincolnshire.

Gloucester Old Spot Sow

Lincoln Curly Coated Sow
Note the similarity in shape, although the
Curly Coat is bigger.

As noted earlier, this breed was bred primarily for its excellent bacon required to feed the many farm workers before the present-day mechanization was developed.

It had the qualities required for the period: large, well-shaped , white, with long curly hair, medium head, neck and ears ; wide, deep chest; broad loin ; large hams and deep sides.

It developed quickly, reaching 30 stone between nine and twelve months old, and up to 40 stone by the twentieth month, which represented a great deal of meat fed from farm produce and that which the pig could gather from the land.

Older generations will remember the fatty bacon from this type of pig and how delicious it was. Nowadays with lectures on cholestrol and obesity there is no demand for this splendid food, more the pity. Nevertheless, there may still be a market for a modified form of Lincolnshire Curly Coat pig and, like the Lincolnshire Buff breed of poultry it might be worth reviving.

Aiming for reasonable weights and selection of breeding stock with less fat, could within a few years produce a pig with desirable qualities -- easy to breed, rear, and fatten -- without the undesirable fat pushing out the lean.

Sadly around 1970 appears to be the last time the breed was to be seen, although, who knows, some remote farmer may still have them on his land, but does not show them or regard them as special.

RECENT HISTORY OF THE PIG

There were many well known breeders of Curly Coats and mention can be made only of one or two. In any case, the Breed Society closed, although there are records consisting of Herd Books and Minute Books which were taken over by the Lincolnshire Archives which is run by Lincolnshire Couny Council.

The Bristow Pigs

A famous strain of Curly Coats was owned by the Bristow family. An extract has been taken from a book on farm stock which recorded as shown below.

The Bnstow pigs were sold not only in Lincolnshire, but were sent to several neighbouring counties. Among their regular customers was Mr G. Sumner, the local blacksmith and Mr Todd of Great Ponton who many will recall as the well-known Lincolnshire Shire Horse man. Local farmers W. Couling of Moulton Eaugate, R. West of Old Leake, J. Woods of Freiston and S. Pearson of Donington were among others who reared the Curly Coats, and Mr James Holmes entered into friendly rivalry with the Bristows when showing his Pinchbeck Herd.

Show days necessitated early rising. By 3am Ernest Bristow was out shampooing the animal with a special shampoo purchased from Osborne's of Grimsby. A brisk rub down with a fine white sawdust completed the operation which usually took the best part of an hour, after which the animal was loaded, in the early days into a Harvey's Cattle Transporter, but later into his own less expensive truck and trailer. Meanwhile, the rest of the herd needed feeding. In the summer months the pigs lived quite happily on grass, raw potatoes and barley meal. In the winter their diet was enriched by the addition of quantities of tick beans, mangolds and the $2^1/_2$ cwt of potatoes which were boiled daily and fed to the herd, then well over 100 strong.

Breeding and rearing were not without their traumas. On one occasion an anthrax scare caused prices and sales to plummet. Bullocks had been found dead in the crew yard, and the local Vet pronounced the dreaded anthrax and refused to call in a second opinion. The following year more bullocks were found dead.

A Spalding vet cut off the ears of the dead animals for an examination which revealed no disease. Thaxters of Moulton, the then local 'cadders', collected the animals and after conducting a post mortem discovered that they had died—fortunately not from anthrax at all—but from a surfeit of cotton cake causing a blockage in the tissues surrounding the heart.

In 1957, himself an appointed Judge of the Curly Coats by now, Ernest Bristow was invited by the Curly Coat Association to enter in yet another Show. Although professing him

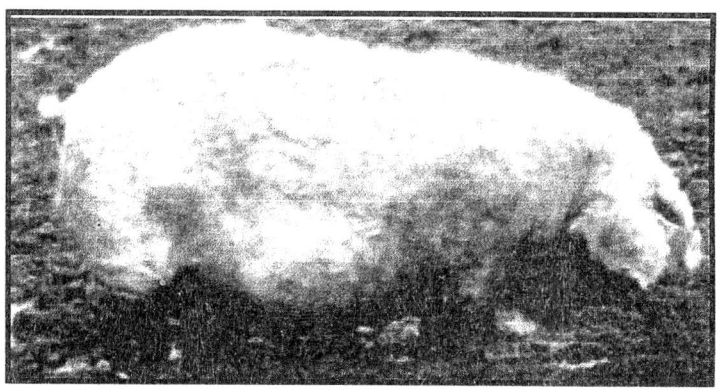

Two Year Old Lincs Curly Coat.

This photograph was taken in 1961 in a very muddy field, with icy sleet, yet this sow was able to cope and thrive, such was the nature of the breed.

self both unready and unprepared, his entry of four took four prizes that year at Brocklesby Park, where for many years Lady Yarborough herself had taken a great interest in the breed.

By the early 1960s the heyday of the Curly Coat was over. and fat was its downfall. The much photographed pig fed to 61 stone by W. Goodger of Boston in 1903 would have been regarded as a monstrosity.

The post-war generations had lost their 'fat tolerance' and the breed fell into decline. By the end of the decade the Lincolnshire Curly Coat was almost extinct and Ernest Bristow had long since decided to call it a day.

Retired now, Mr Bristow's only regret is that his Herd Books are no longer available, having been destroyed on moving to his bungalow home. Among his proudest possessions are souvenirs of many a Show victory—rosettes, spoons and an engraved tankard presented to him in 1956 by the Lincolnshire Agricultural Society.

All these trophies bear witness to the skill of a man whose sound judgement of the pig world would endorse Thomas Tusser's advice as quoted from his 'Five Hundred Points of Good Husbandry' :

In doing of either, let, wit bear a stroke,
For buying or selling of a pig in a poke.

ELDERLY CURLY COAT OWNED BY GEORGE BRISTOW.
He grew fond of the pig and refused to kill it.

This elderly Lincolnshire Curly Coat pig was a favourite of Mr George Bristow. He refused to kill it, and it 'died on the job'. It was an early member of the herd. Breeding improved steadily over the years.

The Lincolnshire Curly-Coated Pig Breeders' Herd Book.

Owner—Mr. W. C. Brown, Appleby, Doncaster.

Willoughbee XII.

H.B. No. 6046.

Farrowed April, 1909.

Breeder—
H. F. Nicholson,
Willoughton Manor.

Sire,
Ruston's
(Lancaster),
H.B. No. 1245,
b H. G. Thorpe.

Dam,
Willoughton
Mary Ann,
H.B. No. 4388,
b F. Nicholson.

Ruston's Chosen,
H.B. No. 726,
b H. G. Thorpe.

Gainsborough's
Masterpiece,
H.B. No. 496,
b T. Ward & Son.

Wrangle Beauty,
H.B. No. 918,
b Wright and
Parker.

Hemswell Count,
475.

Hemswell
Curly-Coat, 458.

Hemswell King, 151.

Ruston's Charmer,
710.

Midville H.B. 221.

Leadenhall Active,
544.

Leake Ajax 107.

Wright & Parker

Hemswell Sam,
H.B. No. 477,
b J. Featherstone

b F. Featherstone.
b F. Smalby.

b W. Beaumont.
b C. Bromby.

Hemswell Duke.
b H. G. Thorpe.

b L. Chevin.
b C. Skinner.

Leadenhall Toby
b T. Ward & Son.

b W. H. Ward.
b J. H. Smith.

Owner—Mr. A. B. Holt, Sturton-by-Scawby, Lincoln.

Holt's Salone.
H.B No. 5786

Holt's Ensles.
H.B. No. 5788.

Holt's Marchioness.
H.B. No. 5790.
Farrowed June, 1910.
Breeder A. B. Holt, Sturton-by-Scawby.

Sire.
Firsby Dreadnought.
H.B. No. 1059.
bJ. H. Smith.

{
  Havenhouse Top Score.
  H.B. No. 464.
  bH. S. Scorer.
  {
    Midville County Councillor, 319.
    { Midville Bob.
    { Midville Bess I.
    East Kirkby Alexandra, 236.
    { Stickney Alex.
    { Stickney Sunbeam.
  }

  Firsby Belle, ...
  H.B. No. 1374
  bJ. H. Smith.
  {
    Firsby Admiral, 441
    { Havenhouse Wep.
    { Heckington A.
    Firsby Abbess, 694.
    { Leake Midville.
    { bWright & Parker.
  }
}

Dam.
Holt's Princess,
H.B. No. 2630.
bA. B. Holt.

{
  Chilebre Moonlight,
  H.B. No. 395.
  bT. Barrand.
  {
    Chilebre Corporal, 393.
    { Chilebre Hove.
    { bT. Barrand.
    Chilebre Buttercup, 90.
  }

  Holt's Wendy,
  H.B. No. 2626.
  bC. P. Barrows.
  {
    Heath George, 130
    { Grange Midville.
    { bG. Free.
    Heath Jane, 560.
    { Heath Tariff Reform.
    { bC. P. Barrows.
  }
}

Class 502. Sow born in 1948.

1st	J. H. Holmes	*Pinchbeck Royal Pride 1st*	Church St, Donington
2nd	J. H. Holmes	*Pinchbeck Royal Pride 2nd*	Church St, Donington
3rd	Reg. W. West	*Leake Countess 3rd*	Ings Farm, Old Leake.
4th	B. Edgar Balderston	*Pinchbeck May*	Spilsby

Pinchbeck – bred by J. T. Larrington, Brothertoft.

Curly Coat Pig : Last Time Exhibited in 1948

Class 502. Sow born in 1948.

1st   J. H. Holmes   Church St., Donington
Pinchbeck Royal Pride 1st

2nd   J. H. Holmes   Church St., Donington
Pinchbeck Royal Pride 2nd

3rd   Reg. W. West   Ings Farm, Old Leake.
Leake Countess 3rd

4th   B. Edgar Balderston   Spilsby
Pinchbeck May

Pinchbeck — bred by J. T. Larrington, Brothertoft.

Curly Coat Pig : Last Time Exhibited in 1948

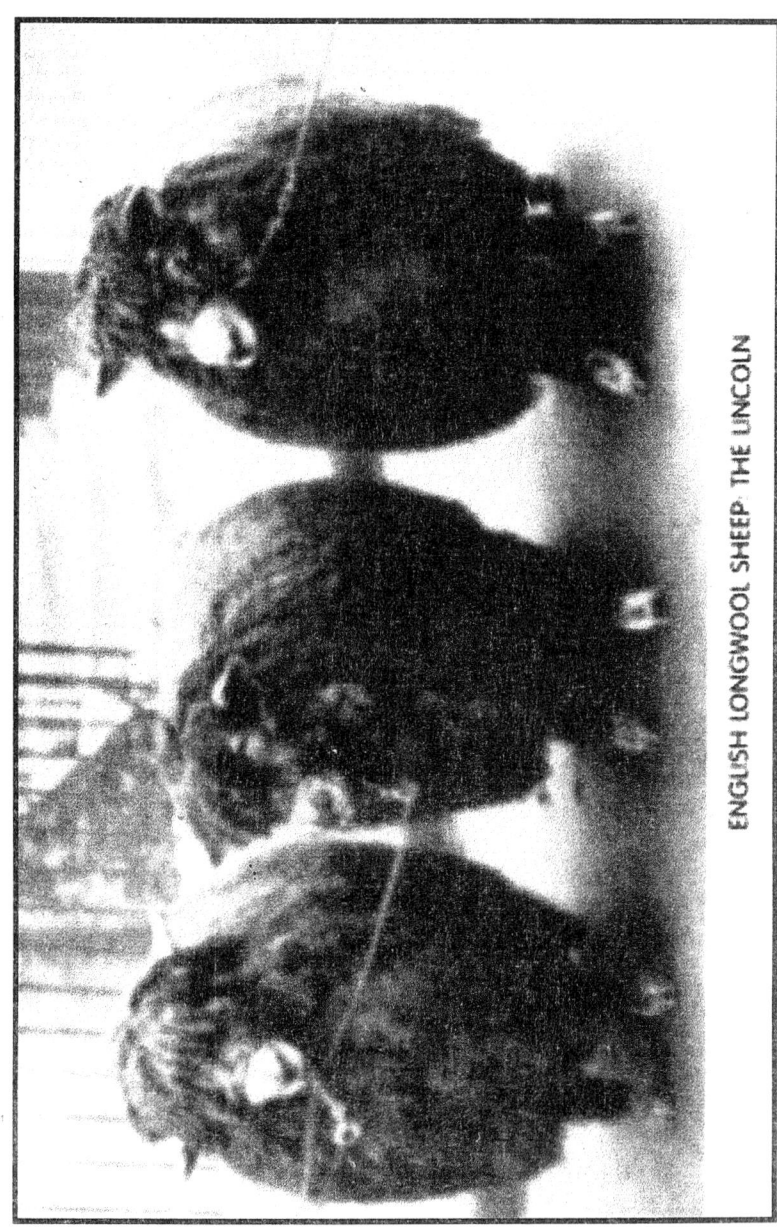

ENGLISH LONGWOOL SHEEP: THE LINCOLN

5

THE
LINCOLNSHIRE
LONGWOOL
SHEEP

LINCOLN SHEEP

Lincoln sheep have a long history and, and initially, were kept for their wool producing qualities, but then were improved so they now excel in meat productionn as well.

HISTORY OF THE LINCOLN LONGWOOL SHEEP
The Original Lincoln

The original breed of Lincoln had a number of problems, although it was the leading breed for wool. Thus it was stated.*

> In many respects the Lincoln is the most important of the Long-woolled breeds of sheep. It is bred pure in greater numbers than any other Long-woolled breed, and its position in the export trade is unequalled by any British breed of sheep. The Lincoln also stands out as the largest and heaviest of all the breeds of sheep, and is unsurpassed in respect of weight of fleece and length of wool.

> It need hardly be said that the breed takes its name from the county of Lincoln, where it appears to have existed for centuries, and where it is still kept in very large numbers. It has extended its range into south-east Yorkshire and to other adjoining counties, but in the main it is still essentially a Lincolnshire breed, and is practically the only breed of sheep bred pure in that famous sheep county.

The Evolution

The breed, as we now know it, has been evolved from the old Lincoln breed by means of a liberal infusion of the blood of the Dishley Leicester.

---

* Professor R G White, Livestock of the Farm, C Brynor Jones, London, nd.

The old Lincoln was pre-eminently a wool-producing animal, the carcass being quite a secondary consideration. The sheep were usually kept until three years old, and occasionally had to be wintered four times before they were fit for market. Culley writes of them as follows:—

> " It is true that the Lincolnshire breeders can justly boast of clipping the greatest weight of wool from a given number of sheep of any other set of people in the island; but then this very heavy wool seldom or never fails to cover a very coarse grained carcass of mutton—a kind of mutton well known for its coarse grain and big bones in the London markets. Yet this is not the worst of it, for this kind of sheep cannot be made fat in a reasonable time."

After quoting cases to show that by crossing with the Leicester a symmetrical and quickly-maturing animal was produced and that though the weight of fleece was reduced this was more compensated for by the fact that the carcass could be marketed earlier, he mentions the reason why for so long Lincolnshire breeders kept to the old breed rather than improve it.

Leicester Ram

Used by Robert Bakewell to improve the utility properties of the Lincoln Sheep, which was developed primarily for its wool, with detriment to other requirements.

From around 1850 changes took place to give a much improved animal.

"The rich fatting marshes in Lincolnshire are, beyond any other county I know of in the island, best adapted to the growing and forcing of good heavy wool. This, with the high price that kind of wool had given previous to the American war, very probably, induced the sheep breeder of that county to pursue it so ardently in preference to every other requisite."

It has also been stated:

These accounts of the origin of the breed are not merely matters of historical interest, but emphasize the fact that, throughout the development of the breed, wool has been regarded at least as much as mutton.

At the beginning of the nineteenth century the practice of crossing with the Leicester became universal, and the modern Liincoln was evolved, being intermediate between the old Lincoln and the Leicester, and retaining the size and fleece of the old breed while acquiring much of the early maturity and symmetry of form of the Leicester.

A typical Lincoln sheep is a very large, deep, and massively-built animal. As an indication of its size, it may be mentioned that the average live weight of shearling wethers, averaging twenty-one and a half months in age, shown at the Smithfield Show, is about 335 lb., that of lambs averaging less than ten months is well over 200 lb.

LINCOLN RAMS

The only other breed which at all approaches these weights is the **Cotswold.**

The carcass is distinguished for its symmetry and good proportions, though the quality of the mutton, as in the case of all long-woolled sheep, leaves much to be desired.   It is apt to contain too large a proportion of fat, sad the mutton is somewhat coarse - grained and deficient in flavour and colour.

The Lincoln is, however, better as regards proportion of lean than some other breeds of the same class. The face and legs of the Lincoln are white; the head is hornless, but well covered with wool, and a tuft grows on the poll and hangs down the forehead.

Cotswold Ram

BRIEF HISTORY

It is reasonable to suppose that the Lincoln longwool breed had its foundation when the national economy was closely tied to the wool trade. In fact, in Norman times it was actively encouraged, by the virtue of the wool trade with the continent and the wealth went towards building towns like Boston and Stamford was produced.

Records indicate that about 1750 the Lincoln sheep was firmly established in the county. This period coincides with the period when Robert Bakewell began his work on constructive sheep breeding. It is probable that Bakewell used Lincoln sheep as his foundation.

In 1796 steps were taken to form a society to safeguard and improve the breed and this body was replaced by the present Lincoln Longwool Breeders Association in 1982. The objects of the association, are the registration of the flocks and rams, the publication of a flock book recording the breeding of the flocks, and to certify the breeding of the many, many thousands of Lincolns which have been sent overseas.

CHARACTERISTICS

The Lincoln is the largest of British longwool breeds. It is a big sheep with great weight and substance. The compact hornless head, has a blue/white face, clear bold eyes, good width between the ears, a wide set of nostrils, long ears pointed

LINCOLN LONGWOOL RAM

slightly forward, and a broad fore lock of wool . It has a wide
level back and well sprung ribs which give great girth .

The sheep is carried on well placed legs showing bone
and strength to carry a big frame freely and easily. The whole
sheep stands square with an appearance of strength, alertness
and character, portraying the great constitution of the breed,
its hardiness and resistance to disease.

Live weight of mature sheep-rams is 250lb.

The Long Wool Fleece

The Lincoln Longwool sheep produces the heaviest and long-
est stapled fleece of any breed in the world. Moreover, the
wool is of uniform quality throughout the fleece,broad and
dense in the staple, showing a fine lustre with a delicate wavy
appearance.

A flock of pure bred Lincoln Longwool sheep including sheep of all ages clips on average of 14 -16 lbs of wool
per sheep and a young ram may easily exceed 30 1b. A young
ewe at first shearing will often clip 25lbs.

The Longwool Carcase

It has a firm well fleshed carcase with a good leg and low wast-
age and it has constantly won many prizes in competition with
other longwool types at the major fatstock shows in this country

82

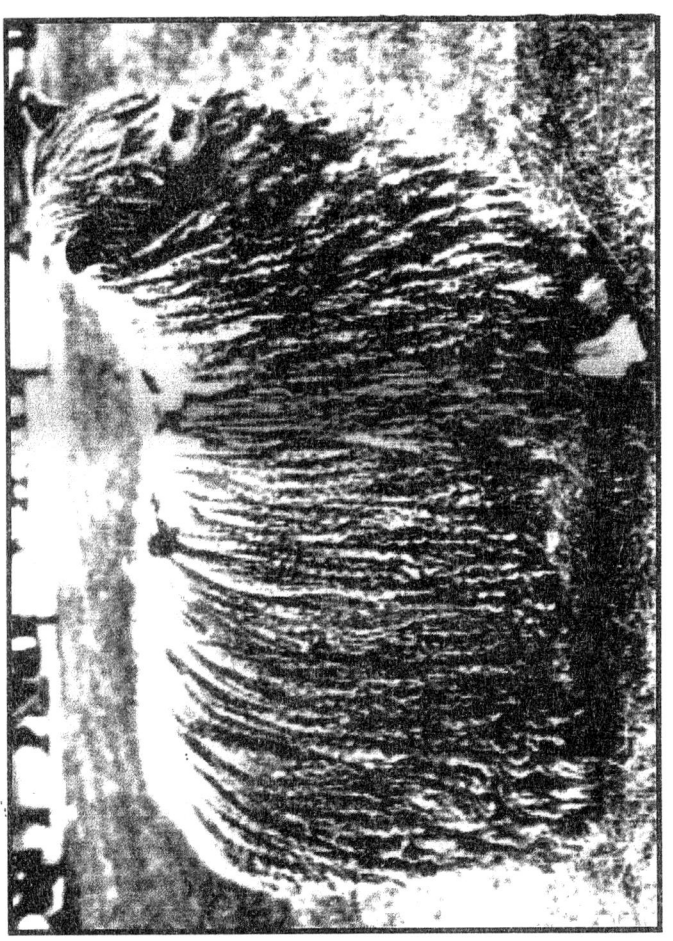

LINCOLN LONGWOOL EWE IN FULL FLEECE

Meat Production

The pure bred Lincoln Longwool lamb can be fattened for slaughter at 7-12 months of age, dependent upon management and the carcase size desired. In England it is customary to slaughter between 9-12 months when the lamb yields a carcase between 60-66 lb dead weight and a substantial fleece.

When fully mature, castrates will fatten to produce heavy mutton carcases. A sheep with the size and wool growth of the Lincoln will not finish at an exceptionally young age, but in the first and second crosses with the smaller short woolled breeds, a lighter weight, quickly finished, can be produced to meet demand for a quality small joint.

Natural Conditions

The home territory of the Lincoln is an area of rolling rather than hill country an area of exposed slopes with little shelter from the persistent cold wind. There is a rain fall average of 26 inches and an ex treme temperature variant of 25-75 degrees F.

The breed lives its whole life outdoors where flocks are folded on grass and arable fodder crops. It is lambed in the spring towards the end of winter under conditions of severely cold winds sometimes snow, despite which little hand feeding is given, under these conditions no breed has been able to supplant the the Lincoln sheep

Lincoln Longwool Records.

A ram sold at a breed sale reached £1,680. A group of 20 rams £4, 500

EXPORT

The largest single purchase a whole flock of 939 sheep excepting a few lambs were all safely delivered to Argentina.

The largest number exported from the United Kingdom in one year was 5,561 and the largest importers before 1940 were Argentina, and after 1940 Russia .

Exports from the United Kingdom have gone on for over a hundred years and large consignments of Lincoln Long wools have been sent all over the world.

Even to day, when disease restrictions impede livestock trade with many countries, exports remain at a remarkably high levels often exceeeding foreign sales of all the other British breeds added together.

Sales to the following countries are recorded by the breeders association:

Argentine and most other South American countries,
Australia,
Belgium,
Bulgaria,
Canada,
China
Czechoslovakia,
Egypt,
Finland,
France,
Germany,
Mexico,
Poland,
Portugal,
Russia,
South Africa,
Spain,
United States of America.,
Yugoslavia..

THE PORTRAIT OF A FAT GIRL.

From a Pencil Drawing by Mrs. Iona Cartwright, wife of a Lincolnshire Farmer

Record of A Fat Cow
Drawing of a Lincolnshire cow from life.
Country Fair, September, 1951.

6

THE
LINCOLN
RED
CATTLE

LINCOLNSHIRE RED SHORTHORN BULL

LINCOLN RED SHORTHORN CATTLE.
History of the Breed Origins

Lincolnshire is an exposed county, which is exposed to the harsh easterly winds, which sweep in from the North Sea.

In past centuries there were invaders who came across that sea from the continent, killing and plundering.

Both factors have left their mark on the county's native live stock which developed in a special way. The Lincolnshire Curly Coat pigs, the Lincoln Longwool sheep, and the Lincoln Red Cattle are all noted for their size and hardiness.

The earliest history of the Lincoln Red breed, with its characteristic cherry redcoat is lost in the mists of time. But there seems little doubt that it has a common ancestry with many of the other large breeds of cattle on the continent of Europe which are today attracting so much interest among cattle men all over the World.

Little is known of the changes, which took place in the centuries that followed, but in 1695 Gervaise Markham in his book *A Way To Get Wealth* (1631 - 38) referred to the large cattle of Lincolnshire which he described as being pied in colour, their horns little and crooked, of bodies exceedingly tall, long and large, lean strong hooved and are indeed fittest to labour and draught.

Developments

During the late 18th and early 19th centuries a number of Lincolnshire breeders brought in good bulls from Durham and York with a view to improving the native cattle. Prominent among these was one Thomas Turnell of Wragby, who is credited with the selection of Durham Shorthorn bulls of an even cherry red colour from the well known herds of Charles Colling and Robert Bakewell, thus founding what was to become known as the Lincoln Red Shorthorn.

ROBERT BAKEWELL
The famous livestock breeder.

CHARLES & ROBERT COLLING
Pioneer Breeders of Livestock

DURHAM CATTLE
Probably now extinct.

RECORDS

The first volume of *Coates' Shorthorn Herd Book*, published in 1822, distinguished between the two types of shorthorn cattle.

In 1896 The Lincoln Red Shorthorn Society published its own herd book for the first time. Initially it was a dual purpose breed, but selective breeding gradually produced two distinct types of Lincoln Red - dairy and beef- and in 1946 the society decided to divide the herd book into two sections, one for entries from beef herds and the other for dairy herds.

The decision was prompted by the differing qualifying standards for dairy and beef bulls imposed by the Ministry o f Agriculture.

A beef breed developed gradually over the years, the emphasis on beef became more marked than that of milk. As a result today the Lincoln Red is a beef breed, though its dual purpose ancestry is reflected in the ample supply of milk which the cows provide for their quick growing calves which mature early.

Not only were they convinced that the trend towards smaller and smaller animals was but a passing phase, but they also knew something that had been noted as long ago as 1799 by Mr Arthur Young then Secretary of the Board of Agriculture. He described the Lincoln as breed of cattle which are unsurpassed in this country for points highly valuable and their disposition at any age to finish rapidly.

.In other words, the fact that the Lincoln Red had traditionally been finished at great weights did not mean that it must necessarily continue to be taken to those weights instead the breeds potential for rapid growth could be exploited in a different way- - by producing cattle for slaughter, but much more quickly than those of the smaller, slower growing breeds.

Perhaps the first sign that the tide was about to turn came with the publication in 1961 of an article in the Ministry's journal *Agriculture* by D L Mason, of the Institute of the Animal Genetics at Edinburgh . This included an analysis of the weight for age of all animals exhibited at The Royal Smithfield Shows between 1950 -- 1959 inclusive. The Lincoln Red entries had the best average weight per - day- of- age of all the breeds.

The Lincoln Red Cow

Not only has the Lincoln Reds steer proved itself as an excellent on range conditions and in the feed lot, but the female is a very desirable suckler cow , which is one important quality that has too often been overlooked in this era of exotic cross breeding.

The Lincoln dam is also a feed-efficient animal and is easily adaptable to regional grazing and climatic conditions, she has a strong broad muzzle and well placed legs. Both characteristics are vital for a good range animal.

Also her feet are sturdy and virtually self maintaining; the udder is neat with well placed teats, enabling the calf to draw its mother's ample milk supply, without any problems.

All these mothering characteristics by themselves seem like small points, but when they are combined with the Lincoln Reds docility and ease of handling, you have a feed-efficient mother that requires very little maintenance or attention.

COLONY CARR HAZEL Winner, owned by H M Prison

This is also shown in the colour section.

THE SHOWS

They also made their mark at the Show and sales winning the championship at Boston and realised the highest price at the Lincoln and Louth bull sales in February 1963. Two interesting figures emerged -- the average price for horned beasts was £132 15s 0d and that for polled £196 17s 6d.

May be the sceptics were happy at last the polls were here to stay. Today the vast majority of Lincoln reds are polled; horned bulls may not be registered.

As long ago as 1960 the breed society dropped the word *Shorthorn* from its title and became **The Lincoln Red Cattle Society.**

97

POLLED HERD AT CROPWELL BUTLER

LINCOLN RED CATTLE ADVANCES

In 1961 the society introduced the first independently observed beef recording scheme which was later to form the basis of the national scheme now operated by the Meat and Live stock commission. As a result they became the first breed society to produce a tangible measure of officially backed performance figures for their cattle and to publish weight figures and gains in their bull sale catalogues.

These performance figures undoubtedly played a major role in attracting the interest of commercial beef producers from abroad to the county breed. It was due to the Lincoln Red Society that the Royal Agricultural Society of England specified that all males entered for the Royal Show classes should be weight recorded.

In 1977 the society took a further step forward by the introduction of the breed development scheme. There was the growing dominance of the continental breeds within the British beef industry and the consumers demand for a leaner product dictated a further need to improve the Lincoln Red.

The selected infusion of continental blood was seen as ideal way of bringing this about to increase the lean meat content of a traditional indigenous breed whist retaining its size and other attributes. Charolais, Maine Anjou, and Limousin and Chianina blood were brought into different herds on a selected sire basis, and the resulting progeny were

BURTON QUALITY III
LINCOLN RED SHORTHORN CHAMPION COW 1908

monitored and selected for further breeding and the results were most rewarding and average back -fat reduced by 50%.

LINCOLN REDS IN THE DAIRY HERD.
Mr John Evens

Mr. John Evens now famous dairy herd at Burton near Lincoln had descended from father to son for generations and he himself tookn it over in 1875.

It was Mr. Evens who has demonstrated to the world, by means of his successes in milking trials and butter tests in England and Ireland, the great dairy properties of the Lincoln Reds.

Mr. Evens commenced showing in 1887, when he carried off the Lord Mayor's champion cup at the London Milking trials with *Beauty*, a cow that gave 3, 673 gal. of milk in thirty four months, and ever since then his showyard record has been one continual series of successes. He has won the first prize and challenge cup at the Belfast milking trials in three consecutive years, and at the Royal Dublin milking trials the first prize and challenge cup in four consecutive years .

At Tring Burton the herd carried off two second , one third and two fourth prizes in seven years, and these are the largest and most representative milking trials in England. One year his winning cow gave 75 lb of milk in twenty four hours and twice his cows have exceeded 71 lb. Besides this, he has repeatedly won at the Oxfordshire Royal Counties, Royal Show, Somersetshire Bath and West, and London Dairy Show milking trials and butter tests, as well as all the leading shows in inspection classes.

Mr G E Sandars

Probably no herd has produced such successful sires in recent years as has the *Scampton* herd belonging to Mr G E Sandars, much of the success of this herd being due to *Keddington Ruby* (1243), whose sons have made more money at the association's sales at Lincoln than have those of any other breeder.

On ten occasions in twenty years Mr. Sandars has secured the highest price for a Lincoln Red and on thirteen occasions he has had the highest average. To him belongs the honour of the record price for a Lincoln Red, that of 305 gns which was paid by Mr Cockbain for *Scampton Goldreef* (4569) a bull destined for Chile.

It is interesting to note that in a dispersal sale catalogue for the Croft herd of Mr J H Searby. *Keddington Ruby* (1243) was born in in 1897, and bred by Mr E H Cartwright. *Keddington Ruby* was used for four seasons in the Croft herd before Mr G. E. Sandars purchased him at the Croft Sale, in 1901 and has since sold 40 bulls by him averaging over 60 gns each, including *Scampton Ex-*

SCAMPTON GOLDLEAF

SCAMPTON EXILE --
LINCOLN RED SHORTHORN CHAMPION

Both are the progeny of *Keddington Ruby*.

pansion (first at Park Royal), and, as noted, ***Scampton Goldreef*** sold for 305gns, and ***Scampton Exile,*** was champion male at the Royal Show 1907 and Lines Show 1908.

FURTHER CHANGES

In 1980 changes were again to take place. No longer viable as a separate entity due to falling numbers in both members, and cattle registered, the Society sought to economise on overheads and join forces elsewhere. Once again joining with the Shorthorn Society was discussed. This had not worked before, and it was decided that their roots should remain within the mantle of The Lincolnshire Agricultural Society.

In effect, this great breed has fallen into decline and steps should be taken to revive before it is too late.

UNREGISTERED STOCK

Although the regidtered stock have declined there are still some herds or single beasts kept by local farmers or small holders. There is no doubting the usefulness of the Lincoln Reds which are docile and easily kept.

The author has an Aunt who still retains a small herd and finds them easy to milk and maintain.

OPPOSITE

The author's aunt with one of her Lincoln Red cows from her small herd of unregistered Lincolnshire Reds. This cow was so

UNREGISTERED LINCOLN RED

Modern Lincolnshire Buffs
Owned by Lucy Hampstead
From *The Lincolnshire Buff Fowl*, Joseph Batty

7

THE
LINCOLNSHIRE
BUFF
POULTRY

Lincolnshire Buff Hen Old Type

Drawn by Harrison Weir

THE LINCOLNSHIRE BUFF FOWL
HISTORY

This was a variety of heavy large fowl, bred in tens of thousands on the Lincolnshire farms, from the 1850's up to the early 1920's and well known in the London markets as the Lincolnshire Buff, but was never registered as a *standard breed*.

A Boston man who wrote to **Kidd's Journal** in 1853 stated , 'no fowl can surpass the Dorking for the table, nor the Hamburgh for eggs'. He also went on to tell of one hatch of half-bred birds, a cross between the Cochin and the Dorking. 'These came off last October, the mother died a fortnight afterwards; however the chicks reared themselves and are now A1.'

He noted that it was very pleasing to watch the little creatures huddle together at night in hay he gave them. He went on "I have not yet tasted the flesh of the thorough- bred birds but those of the cross I speak of are excellent fowls for the table. When six months old they weigh from six to seven pounds each. " (*C P* , Boston , Lincolnshire).

EMERGENCE OF THE LINCOLNSHIRE BUFF

This was the commencement of this particular cross in Lincolnshire, and finding that they were hardy, easy to rear, quick in growth and readily fattened they gradually grew in the

* Most of this chapter is based on the writing of Brian Sands the President of the Lincolnshire Buff Poultry Society.

estimation of farmers and others, and so plentiful did they at last become, that after some few years they were known in the London and other markets as *Lincolnshire Buffs*.

They were thought to have been a cross between Buff Cochin, Red Dorking and the common farm-yard fowl, and possibly Wheaten Old English Game and Gold Hamburgh. The Dorking is one of the oldest of the British breeds and, as a large five-toed dual purpose bird, the Red variety would have been common on the farms in the 1850's. In fact, although rare, it is still around today.

The Buff Cochin originally came to this country from China in the early 1850's where it was known as the *Shanghai* and later as the *Cochin-China*. The breed created a sensation in this country in poultry circles because of its immense size and table properties and excellent egg laying qualities.

Cochin eggs were highly prized in days gone by, and were given to children as a special treat, and were known as Cochin-China eggs. But the Cochins, with their vast amount of feather on body, legs and feet made them unsuitable for the muddy Lincolnshire farms, so this may have been the reason for the development at the end of the 19th century.

In fact, buff coloured poultry remained in favour with the Lincolnshire farmers.

The Beautiful Cochin

Used to develop the Lincolnshire Buff Fowl.

In the early part of the 1890's Mr. William Cook of Orpington in Kent a very famous author and known nation wide for Black and White Orpington poultry, introduced the *Buff Orpington.*

The Controversy*

However, this was thought at the time to have been nothing more than a refined Lincolnshire Buff, and caused a lot of controversy at the time, but with Orpington poultry being famous throughout the country, the Buff Orpington won the day.

Any thing bearing the name Buff Orpington was saleable, or as a Lincolnshire farmer in the *Feathered World* magazine 16th December 1898 wrote:

> "If I call my birds Lincolshire Buff I can't get more than four shillings each for them, but if I call them Buff Orpington they sell readily at ten shillings each."

As late as 1907, Lincolnshire farmers were still advertising in the *Feathered World,* with Lincolnshire Buff in one column and Buff Orpingtons in another, ample proof that they were both bred from the same stock and by 1915 Lincolnshire Buffs had almost disappeared from the Lincolnshire farms.

---

* This and other related matters are considered in *The Lincolnshire Buff Fowl,* Joseph Batty, Elsted, 2007.

The Unique County

Lincolnshire was once unique among the English counties in so much as it had four distinct breeds of farm livestock. We still have Lincoln red cattle and the Lincolnshire long wool sheep, but 1920 saw the demise of the Lincolnshire Buff fowl, but now revived. As recently as 1972 the Lincolnshire Curly Coat pig was declared extinct; these pigs were white and long haired. But as forty stone monsters they fell victim to the modern day trend for a low fat diet.

Red Dorking Poultry
From a painting by Harrison Weir.
They were a major breed used to develop the Lincolnshire Buff Fowl.

THE EARLY SOURCES & A REVIVAL

Fortunately, however, with the help of books published around the turn of the century by authors that include well-known names, such as Lewis Wright, Harrison Weir and Sir Edward Brown, we had at hand information needed to recreate the Lincolnshire Buff.

A project was started in 1981 at the Lincolnshire College of Agriculture and Horticulture, Riseholme in the small animals unit near Lincoln. Unfortunately by 1986 owing to personal changes and Government cut-backs, the unit was closed down and it was at this time that Brian Sands volunteered to take on the project, assisted by other enthusiasts. He as a Lincolnshire "Yellow-Belly" born and bred, and a poultry keeper since a boy, so thought it to be his duty to the county.

He commenced with a remnants of stock from Riseholme, and new infusions of Red Dorking and Buff Cochin blood. There was steady progress over the years, including frequently displaying birds at the Lincolnshire agricultural show.

Then in 1994 Lincolnshire Buffs were shown at the National Championship show as a non-standard breed by Mr Peter Hadgraft and Mr Sands. Considerable interest was shown in the birds and Peter who had recently moved back to his native Lincolnshire and is a man with vast experience in breeding poultry suggested we should form a society to promote the breed.

A meeting was called and a committee formed and on the 29th of January 1995 the Lincolnshire Buff Poultry Society came into being.

A proposed *standard* was drawn up, and the next step was for the breed to be accepted by the Rare Poultry Society, which was achieved in 1996.

This paved the way for the Poultry Club of Great Britain to consider Lincolnshire Buffs as a standard breed, a proposal put before the Poultry Club meeting in January 1997, and they were accepted as a standard breed.

This is probably the first British large fowl to become a standard breed since Reginald Appleyard's White Ixworth in 1932 (excluding the auto sexing breeds) and after a period of almost 150 years the Lincolnshire Buff can now take its rightful place as a true standard British breed.

In this breed we look for a large fowl with males 4 to 5 kg and females 3 to 4 kg, a dual purpose bird keeping as close as possible to the old breed , but to a standard type, with a moderately long body, natural buff colouring (as in Nankin Bantams) with a full copper- bronze tail, (in the male) medium to small single comb, clean white legs and with five toes.

The modern day Lincolnshire Buff is highly suited to today's trend for self sufficiency and organic farming, with its dual purpose qualities, good egg laying, pleasant appearance calm temperament and its ability to brood its own chicks, makes this breed an ideal choice for the small scale poultry keeper.

POULTRY STANDARDS

Standards are written details of a breed. For recognition and registration with the Poultry Club it must be shown that a stable breed has been developed and breeds consistently true in colour and other breed characteristics.

Once approved the breed is given a standard which then allows it to be shown at Poultry Club shows.

STANDARD

LINCOLNSHIRE BUFF FOWL

Origin: British

Classification: Heavy

Egg Colour: Tinted.

Description

A breed found mainly in its native Lincolnshire, but supplied the south of England and London with white fleshed birds in the nineteenth and early twentieth century. Has been redeveloped in the 1980's in Lincolnshire. Are good layers and large, white fleshed birds.

GENERAL CHARACTERISTICS

MALE

CARRIAGE Alert, upright, with bold appearance

TYPE Body large, deep, and moderately long. Back broad, saddle feathers of medium length and abundant.

Breast broad with well rounded keel bone, long and straight. Wings moderately large and carried horizontal.

Tail of medium size and carried well out , with well curved sickles on males.

Head strong, beak stout, eyes large and bright.

Comb, to be single, upright and straight, medium or rather small., free from side sprigs, to be smooth and fine in texture, with five or six evenly serrated spikes.

Face, smooth.

Wattles, medium to small, rounded, smooth and of fine texture. Ear lobes well.developed, pendant fine and rather small.
Neck of medium length with full hackle.

Legs and feet, free from any feathers. Legs set well apart, thighs large, and of medium length.

Toes five, and three front toes lo be large, round, long and straight but well spread. Hind toes double, the fourth toe as near as possible in the natural position, with the fifth toe placed above, with a distinct gap between the two, the fifth toe to be as long a s possible pointing upwards.

FEMALE

The general characteristics are similar lo those of the male, allowing for the natural sexual differences, in that comb, earlobes and wattles are smaller.

COLOUR
MALE PLUMAGE

Back neck and saddle hackles, a rich orange, wing coverts chestnut to copper, Tail, the main feathers black, sickles and side hangers chestnut to copper, shading into black, remainder of the plumage ginger buff to the skin.

FEMALE PLUMAGE

Neck, Back, Wings, Saddle and Tail ginger buff.

Tail shading into black at the end.

Remainder of plumage ginger buff to the skin.

In both sexes, beak buff, eyes bright orange.

Comb, face, lobes and wattles bright red.

Legs and feet white, a small degree of black on inner web of wings.

DEFECTS.

Absence of fifth toe, legs other than white, any feathers on legs,white in face or lobes. Mealiness in surface colour, grey undercolour visible black in the closed wing and any deformity.

SCALE OF POINTS

Type	20
Colour	20
Size	20
Head	15
Feet & Legs	10
Fifth Toe	5
Condition	10
TOTAL	100

WEIGHTS:

Cock	4.00 to 5.00 k. (9 -- 10 lb)
Cockerel	3.10 -- 4.00 k. (7 -- 9lb)
Hen	3.10 -- 4.00 k.
Pullet	2.90 -- 3.60 k. (6 -- 8 lb)

OPPOSITE

In Foreground Trio of Lincolnshire Buff Poultry and behind Toulouse geese. From *The Lincolnshire Buff Fowl*.

The birds owned by Mike Sumner.

OLD TIME CHAMPION LINCOLN RAM, RIBY GLOUCESTER CHAMPION

APPENDIX

BIBLIOGRAPHY

INDEX

BIBLIOGRAPHY

Many titles were studied and, where appropriate, were quoted from.

FARM STOCK OF OLD Sir Walter Gilbey

THE LINCOLNSHIRE BUFF FOWL, Joseph Batty

HARRISON WEIR (Biography) Joseph Batty

ILLUSTRATED BOOK OF POULTRY Lewis Wright

RACES OF DOMESTIC POULTRY Sir Edward Brown

STANDARD CYCLOPEDIA OF MODERN AGRICULTURE

 Prof. R Patrick Wright

VARIATION OF ANIMALS & PLANTS UNDER DOMESTICATION

Charles Darwin

THE ENGLISH COUNTIES Contribution by J Wentworth Day

OUR COUNTRY Cassell & Co.

RURAL LIFE John Sherer

DOMESTIC PIGS H D Richardson

BOOK OF THE FARM H Stephens

MODERN FARMING S Graham Brade Birks

OUR POULTRY & ALL ABOUT THEM Harrison Weir

ELEMENTS OF AGRICULTURE W Fream

HUSBANDRY Walter of Henley

SIX CENTURIES OF WORK & WAGES Thorold Rogers

OLD SPORTS OF THE BRITISH ISLES Hy Alken (BPH)

LIVESTOCK OF THE FARM Prof. R G White

INDEX